by Mary Dylewski

Harcourt
SCHOOL PUBLISHERS

Orlando   Austin   New York   San Diego   Toronto   London

Visit *The Learning Site!*
www.harcourtschool.com

Introduction

Steeplechase Park at Coney Island in Brooklyn, New York, had the world's first roller coaster. The coaster was built in 1884 and was called the Switchback Railway. It had two cars that ran along a wooden track.

The track was about 600 feet long. A tower about 50 feet high stood at each end. The cars started from one of the towers. Gravity pulled the cars downhill. When the cars stopped, the riders got out. Then park workers pulled the cars up to the other tower. The riders walked to the other tower. They climbed stairs to reach the cars. Then they got back in the cars and rode the coaster back!

The Switchback Railway may sound boring. Its hills weren't very high. Its top speed was only 6 miles per hour, but people loved it.

Roller Coasters

The Switchback Railway didn't have a motor or an engine. Today's roller coasters don't have them, either. But today's roller coasters go much faster. For most modern roller coasters, the cars are pulled to the top of the first hill. Just like they were in 1884! Then gravity sends them racing downhill.

So why do they go so much faster today? Today's roller coasters start from much higher hills. The Switchback was 50 feet high. Today's coasters are often more than 200 feet high.

Super Swings!

Some rides are like big swings. You move back and forth in a big arc. The higher you start on one end, the higher you go on the other end. If you continued all the way around, you would go in a big circle.

Gravity is at work in this ride, too. When you start to swing, gravity pulls you down toward the ground. Your motion keeps you swinging up the other side. Then you glide back toward the ground, swing up backward, and start over.

When you are moving down, you move fast. When you swing up, you move more slowly. Big swing rides have motors that keep them going.

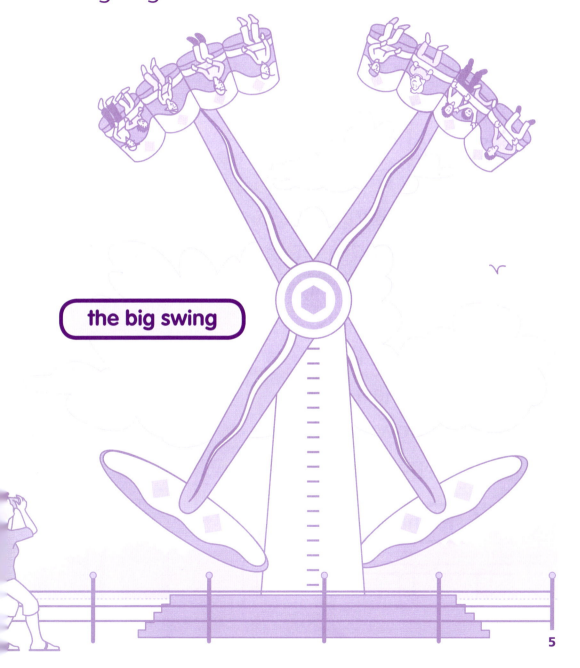

the big swing

Free-Fall Rides

Gravity also provides the fun on free-fall rides. On a free-fall ride, riders are pulled to the top of a tower. Then the track goes straight down. Riders drop. Gravity pulls them down fast! Most free-fall rides let the riders fall for only one or two seconds. Then the track turns straight and level. The cars slow down and gently stop.

(falling free)

Carousels

Brightly painted horses gallop in a circle on a carousel. Some of the horses on a carousel are near the center of the circle, and some are near the outside.

The horses on the outside move in bigger circles than the horses on the inside do. All of the horses go around the circle in the same amount of time. This means that the outside horses move faster than the inside horses.

Bumper Cars

In bumper cars, you take control. You drive your bumper car around and bump other cars. You can go forward, backward, and around in circles. The cars have big rubber bumpers so you don't bump too hard.

The faster you go, the harder you bump other cars. When cars bump, they bounce off each other and change directions.

In bumper cars, you get to drive. ▼